Vanderléia de Castro Bezerra
Fabiola Souza
Kemislani Lima

Use of WTP sludge in clay for ceramic brick production

Vanderléia de Castro Bezerra
Fabiola Souza
Kemislani Lima

Use of WTP sludge in clay for ceramic brick production

An environmentally friendly solution for this waste: WTP sludge

ScienciaScripts

Imprint

Any brand names and product names mentioned in this book are subject to trademark, brand or patent protection and are trademarks or registered trademarks of their respective holders. The use of brand names, product names, common names, trade names, product descriptions etc. even without a particular marking in this work is in no way to be construed to mean that such names may be regarded as unrestricted in respect of trademark and brand protection legislation and could thus be used by anyone.

Cover image: www.ingimage.com

This book is a translation from the original published under ISBN 978-613-9-69698-7.

Publisher:
Sciencia Scripts
is a trademark of
Dodo Books Indian Ocean Ltd. and OmniScriptum S.R.L publishing group

120 High Road, East Finchley, London, N2 9ED, United Kingdom
Str. Armeneasca 28/1, office 1, Chisinau MD-2012, Republic of Moldova, Europe
Printed at: see last page
ISBN: 978-620-8-13382-5

SUMMARY

Water treatment plants transform raw water into potable water for human consumption through conventional methods of coagulation, flocculation, decantation, filtration and disinfection. During this purification process, solid waste is generated. This by-product, which comes from the addition of chemical substances, is called sludge and is self-degrading, with the potential for negative impacts which, if released into the environment, cause serious damage to the soil and aquatic life. The aim of this study is to assess the potential use of WTP sludge by incorporating it into clay for making ceramic bricks, in order to present an alternative environmentally correct way of disposing of this waste. For this purpose, laboratory tests were carried out with proportions of 0%, 12%, 15% and 20% of waste incorporated into the clay mix, the results of which could influence the physical and mechanical properties of the ceramic material. Chemical and mineralogical characterisation was carried out using X-ray fluorescence and X-ray diffraction, respectively, to quantify the elements present in the raw materials. Physical and mechanical tests were carried out to assess the quality of the final product by analysing the properties of linear shrinkage, water absorption, apparent porosity, apparent specific mass and flexural tensile strength. By comparing the results with the parameters established in NBR 15270-1, 2 and 3/2005, it was proved that WTP sludge can be incorporated up to 20% into the clay mix for brick manufacture.

KEY WORDS: WTP sludge, clay, pollution, ceramic brick

SUMMARY

CHAPTER 1

INTRODUCTION

With the increase in globalisation and the corresponding rise in water consumption, people have started to look for new ways of obtaining water in good conditions for use (TAKADA et al., 2013, p. 158). Public water treatment plants (WTPs) transform water that is unsuitable for human consumption into water that is sanitary and safe, in accordance with the potability standard established in Brazil by the Ministry of Health's Ordinance No. 2,914/11 (TSUTIYA et al., 2001, p. 1).

The water that arrives at the Treatment Plant, due to its physical, chemical and biological impurities, needs to go through a purification process to remove pathogenic microorganisms and contaminating components, which requires the addition of chemical products such as aluminium sulphate, lime, chlorine and polymers.

When the water supply industry uses complete or conventional treatment (coagulation, flocculation, decantation and filtration), it carries out processes and operations such as the introduction of chemical products, which generate waste (REIS et al., 2006, p. 210).

This by-product, generated by the addition of chemical products and water, is called sludge and is basically made up of soil particles, organic material carried into the raw water. It is classified as waste by the Brazilian Technical Standards Agency (ABNT) 10.004/2004, which establishes solid waste management procedures.

According to Castro et al. (2015, p. 47), the sludge generated at WTPs has an inadequate final destination and is exposed to the environment, contaminating it. According to Silva et al. (2013, p. 399), soil decomposition and contamination of water sources and groundwater are examples of environmental damage caused by improper disposal of waste.

This material is a compound capable of causing environmental pollution because it contains chemical substances that, if disposed of in the environment without proper treatment, can cause serious environmental damage to soil and aquatic life.

In Brazil, the National Solid Waste Policy, Law No. 12.305 (Brazil, 2010), prohibits the dumping of waste in natura in the environment and refers to the need to reuse waste resulting from human activities, which must be disposed of properly (BILDHAUER, D. C. et al.2015, p.74).

The texts of the National Water Resources Policy - Law No. 9.433/97 and the Environmental Crimes Law - Law No. 9.605/98, make it an environmental offence to dispose of WTP sludge in natura, but according to Tsutiya (2001 p. 1) sludge is often disposed of in waters close to the stations.

The amount of waste generated at WTPs is very large. In Brazil, in the Greater Vitória region, it is 350 tonnes/day, and the expectation is that this figure will grow even more as the population joins the sewage collection network.

According to Jimenez (2011, p. 15), in order to overcome the environmental problems caused by sludge, it is necessary to develop new materials and look for new construction solutions, such as reuse and recycling systems. Ramos (2015, p. 15) states that a study is needed to enable WTP waste to be used in the ceramics industry.

One of the waste products that has the potential to be recycled into red ceramics is the sludge generated by water treatment plants (PINHEIRO et al., 2014, p. 205). The physical and chemical characteristics of WTP sludge are often similar to those of the materials used to make bricks: clay (KATAYAMA, 2012, p.13).

The existing deposit of clay raw materials is extensive in the region of Cacau Pirera, a district in the municipality of Iranduba, and contributes to the expansion of the pottery industry. The most widely used raw material for making bricks is clay.

Pinto and Pinheiro (2013, p. 436) state that oil production depends directly on natural resources, especially clay and wood. The exploitation of the clay deposit at the Iranduba centre, on the right bank of the River Negro, is considered negative due to its environmental impacts. According to Pinto and Pinheiro (2013, p. 436), the impacts caused by the pottery industry include the opening of ditches to extract the clay, deforestation to remove the wood and the production of smoke.

Both the extraction of clay and the generation of WTP sludge cause serious environmental problems for water, soil and air. Despite this, Aquino et al (2015, p. 30) states that the problem of sludge disposal would be solved if the sludge were successfully incorporated into clay.

The incorporation of sludge into ceramic material helps to reduce pollution and, at the same time, provides an alternative to brick production in potteries (RAMOS 2015, p. 16).

The motivation behind this work is to help reduce the mineral extraction of clay, as well as to reduce the dumping of sludge in rubbish dumps and bodies of water, which cause environmental impacts and public health problems.

Based on this principle, this study aims to use water treatment plant sludge in clay for the production of ceramic bricks in proportions of 12%, 15% and 20%, in order to make an alternative final disposal of the waste viable.

CHAPTER 2

OBJECTIVES

2.1　　　　　General

To assess the potential use of water treatment plant sludge by incorporating it into clay to make ceramic bricks.

2.2　　　　　Specific

C Characterise the clay and sludge from the Water Treatment Plant (WTP) using laboratory tests;

A To analyse the physical characteristics and mechanical properties after adding water treatment plant waste in proportions of 0%, 12%, 15% and 20%.

CHAPTER 3

MATERIAL AND METHODS

This work was based on an experimental scientific approach. Gil (2002, p. 163) argues that an experiment is the best example of scientific research.

Two raw materials were used in this study: water treatment plant (WTP) sludge and clay. The first raw material, clay, was collected during field research at the Montemar ceramics company, located at kilometre 30 of the Manuel Urbano highway, in the state of Amazonas, in the municipality of Iranduba. The second raw material, sludge, was collected from the Water Treatment Plant (WTP) of the Water for Manaus Programme (PROAMA). Figure 1 shows, in simplified form, the experimental procedure used.

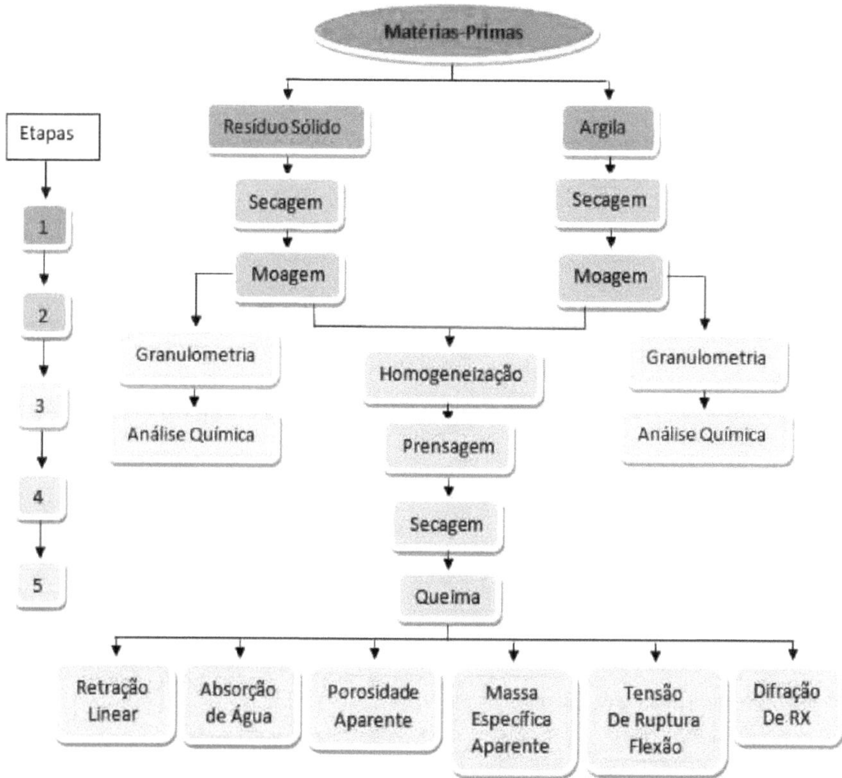

Figure 1 - Flowchart of the procedures and analyses carried out. Adapted Source: Jimenez, 2011.

The study of stage 1 began with the collection of soil samples (clay) and organic waste (sludge). Then, in stage 2, the materials were prepared by drying (clay) and preparing the sample for physical tests, according to NBR ABNT 6457/86 - Soil samples - Preparation for compaction tests and characterisation tests. And also by grinding (silt).

Subsequently, in stage 3, the two materials were characterised using a particle size test in accordance with ABNT standard NBR 7181/82 - Soils - Particle size analysis.

The sieving process and sieve specifications are in accordance with the ABNT NBR NM-ISO 2395/ 97 standard - Test sieves and sieving tests - Vocabulary. Also included in this analysis of physical indices was the actual grain density of the clay and silt, in accordance

9

with standard DNER-ME 093/94. Finally, the chemical composition was analysed using X-ray fluorescence.

In stage 4, the samples were prepared to make the specimens by drying, grinding, homogenising, pressing, drying and firing.

The waste was mixed into the clay mix in proportions of 0%, 12%, 15% and 20%. In the final stage, in order to study the physical and mechanical properties of the test specimens after firing, calculations were made for linear shrinkage (LR), water absorption (WA), apparent specific mass (ASM), flexural tensile strength (FRS) and X-ray diffraction (XRD).

3.1 Collection of raw materials.

3.1.1 WTP organic waste (sludge)

Sludge is generated in the conventional water treatment process for urban supply, in the decantation and filter cleaning units of the WTP.

The waste was collected from the Water for Manaus Programme Station - PROAMA, located in the Armando Mendes neighbourhood, in the east of the city of Manaus (Figure 2). It has the capacity to treat 210,000 litres of water a day and serves all the neighbourhoods in the north and east, benefiting around 500,000 inhabitants.

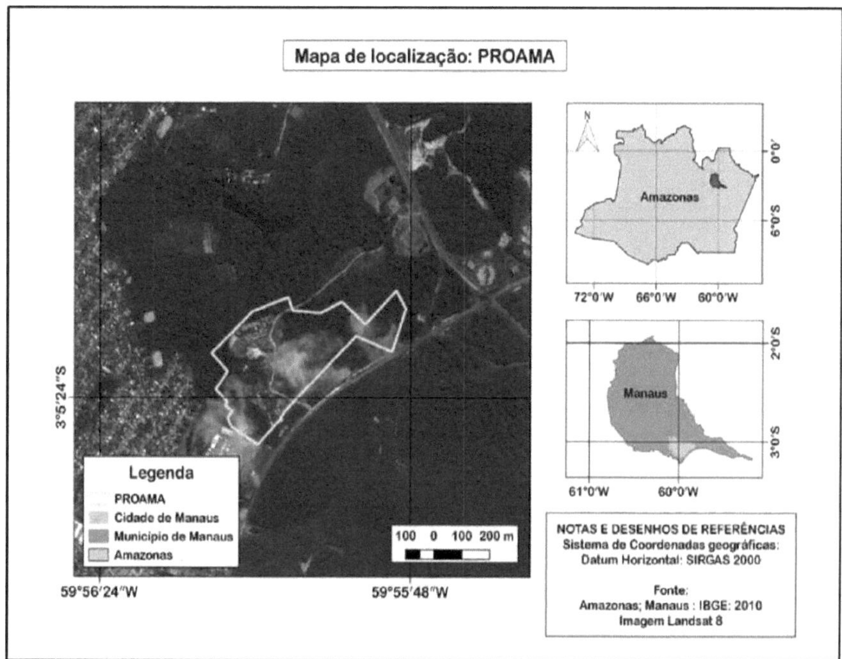

Figure 2 - PROAMA location map. Landsat 8 image. Source: IBGE, 2010.

On the map of the state of Amazonas, there is a red dot, which in the figure below shows the location of the city of Manaus and on it, the yellow dot, which shows the PROAMA area in the yellow circularity.

The waste was packed in a jute fibre bag lined with polyethylene and taken to the Soil Mechanics and Geotechnics laboratory at the Centro Universitário do Norte - Uninorte. It was dried, pulverised, ground and tested for moisture, particle size and density.

Figure 3 a and b shows the residue in its raw state and after grinding the raw material and sieving it, respectively.

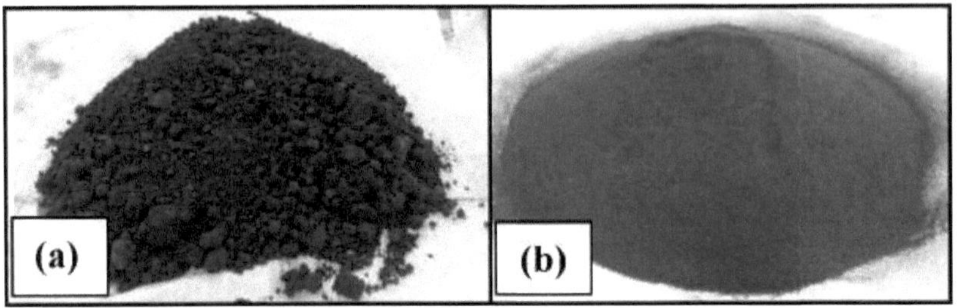

Figure 3 - Raw residue (a), ground and sieved residue (b).

3.1. 2Clay

Figure 4 shows the location of the Montemar Pottery from where the clay was collected.

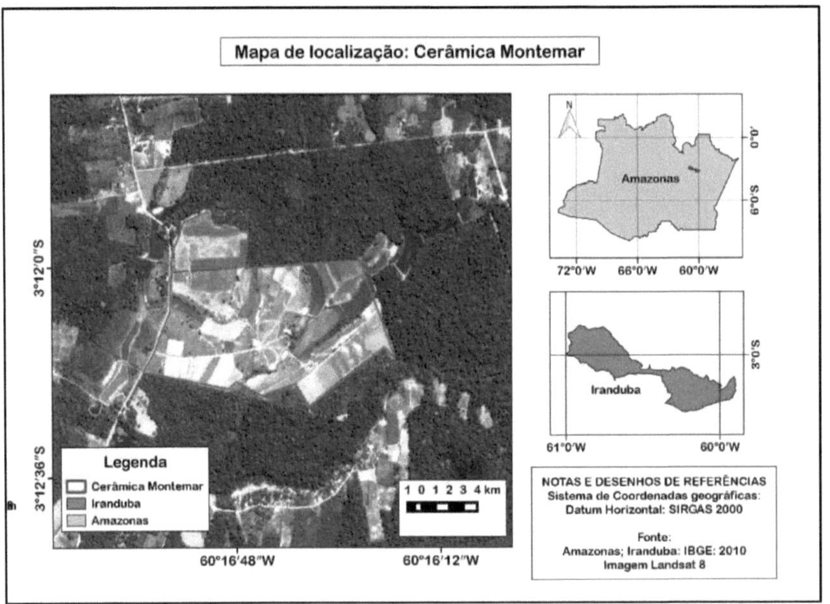

Figure 4 - Location map of the Montemar Ceramic Pottery. Source: Landsat 8 image. IBGE, 2010.

The soil sample was removed from the ground using a mechanical system with the aid of a backhoe and bucket.

The material was packed in cardboard tubs sealed with steel ties and taken to the Soil Mechanics and Geotechnics laboratory at the Centro Universitário do Norte - Uninorte, where the grains were broken up and the material homogenised to determine the moisture content, particle size analysis and the actual density of the grains.

3.2 Characterisation of raw materials.
3.2. 1 Sample preparation for characterisation tests

The samples collected (clay and solid waste - sludge) were dried in the open air until they were close to hygroscopic humidity, on a tarpaulin.

The standard used for this test method was ABNT NBR 6457/86 - Soil samples - Preparation for compaction tests and characterisation tests. It involves breaking up the clods, taking care not to break the grains, homogenising the sample and passing it through the %" splitter to reduce the sample, in order to obtain a more representative sample in sufficient quantity to carry out the test.

After preparing the samples, 2000g of each sample was weighed on a precision scale and passed through a No. 10 sieve with a mesh opening of 2.00mm. The material retained on the sieve was identified as coarse material. Of the material that passed through, 200g was weighed and identified as fine material (Figure 5).

Figure 5 - Residue retained on the sieves.

The materials were then washed under running water. The coarse material was washed through a No. 10 sieve with a mesh size of 2.00mm and the fine material through a No. 20 sieve with a mesh size of 0.075mm, so that no material was lost during the washing process. This process of washing the material is carried out in order to remove all the clay coated on the grains.

The samples were placed in a metal container and placed in an electric oven, automatically controlled by its thermostat at a temperature of 110 C°, for 24 hours.

After this period, the coarse and fine materials were sieved using a sequence of sieves forming a set specified in accordance with the ABNT NBR NM-ISO 2395/97 standard - Test sieves and sieving tests - Vocabulary. Table 1 shows the classification of soils according to grain size, in accordance with ABNT NBR 6502/95.

Table 1 - Soil classification.

Type of soil	Grain diameter (mm)
Clay	< 0,002
Silt	0,06- 0,002
Sand	2,0- 0,06
Pebbles	>2,0

Source: ABNT NBR 6502/95

It can be seen that the diameter of the grains in the clay soil is less than 0.002mm.

3. 2.2 - Sieving particle size test

The sieving granulometry test is standardised by ABNT NBR 7181/ 82 - Soils - Granulometric analysis.

The particle size is determined through a sequence of sieves in ascending order, i.e. from the closest mesh opening to the most open mesh.

The material removed from the oven was passed through this set of sieves and sieved. The material retained on each sieve was weighed on a precision scale and the information obtained was recorded.

The following expressions were used to obtain the results in % of sieving:

M Total mass of the dry sample

$$Ms = \frac{(Mt-Mg)}{(180+h)} \: x \: 100 + Mg \qquad (1)$$

Being:

Ms = total mass of the dry sample

Mt = total mass of air-dried sample

Mg = mass of dry material retained on the 2.0mm sieve

h = hygroscopic humidity of the material passed through the 2mm sieve.

P Percentage of materials passing the 16, 30, 40, 50, 100 sieves.

$$Qg = \frac{(Ms-Mi)}{Ms} x \ 100 \qquad (2)$$

Being:

Qg = percentage of material passed through each sieve

Ms = total mass of the dry sample

Mi = mass of retained material accumulated in each sieve

1.1.3 Determining the moisture content of soil samples.

There are two standards that prescribe this test method: ABNT NBR 6457/ 86 - Soil samples - Preparation for compaction tests and characterisation tests, where there is a brief but also essential description of the analysis, and DNER - ME 213/ 94 - Soils - Determination of moisture content, where there are more details on the procedures.

To check the moisture content of the soil, 3 aluminium capsules numbered 03, 50 and 52 were weighed on a balance sensitive to 0.1g, a small amount of moist material was added and weighed again. The capsules were then taken to an oven automatically controlled at 110°C by its thermostat for 24 hours, with the temperature controlled at 110°C.

in order to completely eliminate all the water retained in it. After drying, the material was weighed.

To obtain the moisture content, the following expression was used:

$$h = \frac{M_1-M_2}{M_2-M_3} \ X \ 100 \qquad (3)$$

Being:

h = moisture content, in %;

M1 = mass of moist soil plus mass of container, in g

M2 = mass of dry soil plus mass of container, in g

M3 = mass of the container (metal capsule with lid) in g

1.1.4   Real Grain Density (heated pycnometer)

The tests were carried out on samples of clay and organic waste (sludge) separately.

To determine the actual density of the grains, a 50ml pycnometer with an emery stopper was used. The aim is to eliminate the air in the microparticles of the samples being analysed.

The standard that prescribes the test method is DNER-ME 093/ 94 - Soils - Determination of actual density.

Figures 6a and 6b show the test to obtain the density of the residue and show the agitation of the equipment, respectively.

Figure 6 - Pycnometer heating (a), Pycnometer stirring (b).

Firstly, the empty pycnometer (P_1) was weighed and 20g of the sample was inserted inside (P_2). The material was covered excessively with distilled water and the pycnometer was heated with the aid of an asbestos screen to avoid direct contact between the flame and the glassware, leaving the water to boil for 15 minutes. While the water was boiling, the pycnometer was shaken sporadically to avoid overheating. The purpose of heating is to expel all the air between the microparticles in the samples.

After the above-mentioned period, the equipment was left to cool at room temperature. Distilled water was then added up to the neck of the pycnometer and it was placed in the water bath at room temperature for a further 15 minutes, and the bath temperature (t) was recorded. The pycnometer was then removed from the bath and wiped dry with a clean, dry cloth. The pycnometer plus the sample (P_3) was weighed. All the material was then removed from the equipment. The pycnometer was rinsed and completely filled with distilled water and weighed (P_4).

The following formula was used to obtain the actual density of the grains;

$$Dt = \frac{P2 - P1}{(P4 - P1) - (P3 - P2)} \tag{4}$$

Being:

Dt - actual soil density at temperature t;

P1 - weight of the empty, dry pycnometer, in g;

P2 - weight of the pycnometer plus the sample, in g;

P3 - weight of pycnometer plus sample plus water, in g;

P4 - weight of the pycnometer plus water, in g.

3.2.5 X-Ray Fluorescence (XRF)

The chemical analyses of the clay and silt and X-ray fluorescence were carried out in the Chemistry laboratory of the Crowfoot Group at the School of Technology of the Amazonas State University.

This is a multi-elemental, non-destructive analytical technique used to obtain quantitative and qualitative information on the elemental composition of samples.

To analyse the samples, the organic waste (sludge) and clay were formed into boric acid tablets and pressed in a RIGAKU model PCA 30 pneumatic press. The samples were then placed in a RIGAKU Supemini X-ray fluorescence spectrometer in order to analyse the chemical elements present in the samples.

3.2.6 X-ray diffraction (XRD)

The X-ray diffraction analyses were carried out in the mineralogy laboratory of the Geoscience Department of the Federal University of Amazonas (UFAM). In order to analyse the structure and identify the mineralogical composition of the clay and the WTP sludge, an

X-ray diffraction apparatus brand LAB-X SHIMADZU, model XRD-6000 with a ceramic X-ray tube and copper anode, was used.

The instrumental conditions used were: 5° to 7° at 2Θ; 40 KV voltage and 30 mA current; step size: 0.017° at 2Θ and 10.34s time/step; 1/4° divergent slit and 1/2° pre-spread; 10mm mask; sample in circular motion with a frequency of 1 rev/s. The software used for data processing was Origem Pro 8.

3.3 Preparation   of the specimens

The preparation of the samples, as well as the formulation, homogenisation, pressing, drying and physical and mechanical testing of the specimens were carried out in the Geochemistry laboratory of the Geoscience Department of the Federal University of Amazonas (UFAM). CPs were made with a mixture of waste and clay in proportions of 12%, 15% and 20%, for analysis after drying and firing, following the steps of homogenisation, pressing, drying and firing, as shown in the flowchart (Figure 1). CP was also made with 0% waste to serve as a parameter for comparing the percentages.

3.3. 1 Sample preparation

The samples were dried and sterilised in a MARCA Marconi MA033 oven at 110°C for 24 hours to remove moisture, in the Geoscience laboratory at the Federal University of Amazonas (UFAM). The materials were then ground (Figure 7).

Figure 7 - Ball mill.

The equipment used a 10-litre QUIMIS chiarioiti ceramic pot, the purpose of which is to break down and reduce the size of the grains. The clay and waste were ground separately for 2 hours. The material is placed inside the ceramic pot and the pot is placed under the equipment. When activated, it causes the

The mill makes circular movements and the balls contained inside the device break up the material.

Each sample was then passed through a 120mm mesh sieve. The material that passed through was used to formulate the ceramic mass and make the test specimens.

3.3.2 Formulation and homogenisation

To formulate and homogenise the ceramic mass, an Adventurer OHUS model ERC Class II analytical balance was used, calibrated in accordance with Ordinance No. 002/2004. Figure 8 shows the weighing of the raw material with the residue.

Figure 8 - Weighing the raw material with residue.

At this stage, the raw material was weighed, varying the concentration of residue by 0%, 12%, 15% and 20%, with a humidity of 10%, to facilitate shaping during the pressing of the specimens, as shown in Table 2.

Table 2 - Formulation and homogenisation of the ceramic mass

Specimens	Ceramic putty	% by weight of raw materials			
		Clay	Waste	Humidity	Total
1	0%	90	0	10	100
2	12%	78	12	10	100
3	15%	75	15	10	100
4	20%	70	20	10	100

3.3. 3Pressing

To obtain the specimens, the ceramic mass went through the pressing process, at which stage the structural shape of the piece was obtained (Figure 9).

Figure 9- Press (a), Pressing (b), Removal of the specimen.

Five (5) specimens were made for each percentage of waste, totalling twenty (20). To achieve the desired structural shape, the mass was placed in the mould and taken to a MARCONI hydraulic press with a maximum capacity of 2 tonnes under a pressure of 25MPa. The specimens were then ejected from the mould using the same press.

Then, using a 200mm/8 precision digital analogue caliper, ZAAS Precision, Stainless Hafoi model, graduation 1/129" 0.05" mm, accuracy ±0.05, the specimens were measured to determine length, width and thickness, respectively. The dimensions were 72.03 x 31.67 x 7.77 mm, with an average weight of 29.57g, obtained using an Adventurer OHUS analytical balance, model AR 3130 Class II, with a capacity of 0.001g to 310g, calibrated in accordance with INMETRO/DIMEL regulation 002/2004. The force of gravity was disregarded in order to determine the green density of the material.

Measurements and weighing are essential to obtain mass and volume, variables that will be used to determine density, using the formula:

$$D = \frac{m}{v} \tag{5}$$

Being:

D= Density (g/mm³);

m= Mass;

v= Volume

1.1.5 Drying the specimens

The aim of drying was to remove moisture from the surface of the test specimens by evaporation and by diffusion on the inside. To avoid accelerated moisture loss, care was taken to control the drying temperature, as a negative effect of these contractions can occur when the evaporation rate is higher than the diffusion rate, the surface will dry out, contracting its volume faster than the inside, causing deformations in the samples.

The test specimens were dried in a MARCONI MA033 oven, FO7170 series, at a temperature of 110°C for 24 hours.

After this period, the samples were weighed and sized to check the expansion or shrinkage of the minerals exposed to this temperature.

1.1.6 Firing the specimens

The test specimens were fired at the Institutional Thematic Laboratory for the Chemistry of Natural Products (LTQPN) of the National Institute for Amazonian Research (INPA).

The specimens - CP - were taken to the Chemistry of Natural Products Institutional Thematic Laboratory - LTQPN, at the National Institute for Amazonian Research - INPA, where they were fired in a QUIMIS microprocessor muffle furnace. Table 3 shows the temperature variations during firing, as well as the thresholds.

Table 3 - Firing temperature of the specimens

% by weight of raw materials		Specimens	Firing temperature (°C)			
Clay	Waste	Quantity	Start Time	Temperature variation 5min.	Temperature variation 5min.	Final temperature variation 30min.
90	0	5	30°C	450°C	600°C	900°C
78	12	5	30°C	450°C	600°C	900°C
75	15	5	30°C	450°C	600°C	900°C
70	20	5	30°C	450°C	600°C	900°C

The sintering temperature was 900°C, with (three) different stages, 1ª (first) from 0:05min. to 450°C, 2ª (second) from 0:05min. to 600°C and the final variation from 0:30min. to 900°C, totalling a firing time of 2h:40min (two hours and forty minutes). The initial temperature was 30°C and the cooling rate was 75°C per hour over a 12:00 (twelve hour) interval.

After firing, the masses of the specimens were weighed on an Adventurer OHUS analytical balance, model AR 3130 Class II, with a capacity of 0.001g to 310g, calibrated in accordance with INMETRO/DIMEL Ordinance No. 002/2004, as well as the length, width and thickness, width and thickness, measured using a 200mm/8 precision digital analogue caliper, ZAAS Precision, Stainless Hafoi model, graduation 1/129" 0.05" mm, accuracy ±0.05, in order to analyse and check linear shrinkage respectively.

3.4 Physical and mechanical tests

Through physical and mechanical tests, the properties of the ceramic are assessed, which is of paramount importance as it determines the quality of the product, whose parameters are established by NBR's 15.270-1, 15.270-2, 15.270-3. The following tests were carried out:

-Pressing humidity (UP);

- Fire loss (FP);

- Linear shrinkage after firing (LR);

- Water Absorption (AA);

- Apparent Porosity (AP);

- Apparent Specific Mass (ASM);

- Flexural tensile strength (TRF);

- X-ray diffraction (XRD)

3.4.1 Pressing Moisture (UP)

Pressing moisture was obtained using the measured values of the wet mass and dry mass of each sample, taken before and after drying, respectively. The pressing moisture value is calculated using the equation:

$$U\ (\%) = \frac{(Mu - Ms)}{Ms} x100\ (g) \qquad (6)$$

Being:

U= Humidity;

Mu= Wet Mass;

Ms= Dry mass

3.4.2 Fire loss (FP)

To calculate the percentage change in mass after firing at 900°C, the loss-on-fire equation was used.

$$PF\ (\%) = \frac{(Ms - Mq)}{Ms} x\ 100\ (g) \qquad (7)$$

Being:

PF= Fire Loss;

Ms= Dry mass;

Mq= Mass after burning

3.4. 3Linear Retraction   (LR)

The lengths of the specimens were measured before drying and after firing, where linear shrinkage was determined using the equation:

$$RL\ (\%) = \frac{L0 - L1}{L0} x100\ (mm) \hspace{2cm} (8)$$

Being:

L0= Green length;

L1= Length after firing

3.4.   4Water absorption   (AA)

The water absorption test is standardised by NBR 15.270-3. The specimens were placed in a container, which was filled with enough water at room temperature to keep them completely immersed (Figure 10).

Figure 9 - Water absorption.

After gradually heating the container, the water inside boiled and remained boiling for 2 hours. The volume of water evaporated from the container was replaced so that the immersion of the specimens would not be compromised. After this period, the specimens were cooled by slowly replacing the hot water with water at room temperature.

Once the water in the container was at room temperature, the saturated specimens were removed and placed on the bench to allow the excess water to drain off.

The remaining water was removed using a clean, damp cloth, making sure that the time elapsed between the removal of excess water from the surface and the end of weighing was no less than 15 minutes, as established by the standard. After this process, the samples were re-measured.

To calculate the water absorption of the samples, the equation used was:

$$AA\ (\%) = \frac{Pu - Ps}{Ps} x100\ (g) \qquad\qquad (9)$$

Being:

AA = Water Absorption;

Pu = mass of the wet body (g);

Ps = dry body mass (g)

3.4.5 . Immersed mass (IM)

Each specimen had its mass measured when immersed in water. The time the water was left to boil was enough for the water to penetrate the open pores formed by burning the specimens. The values from this measurement were used to calculate apparent porosity and apparent specific mass.

3.4.6 . Apparent Porosity (AP)

The apparent porosity of the samples was determined using the following equation:

$$PA\ (\%) = \frac{Pu - Ps}{Pu - Pi} x100\ (g) \qquad\qquad (10)$$

Being:

PA = Apparent Porosity;

Pi = mass of the body immersed in water (g)

Pu = mass of the wet body;

Ps = dry body mass

3.4.7 . Apparent specific mass (ASM)

The apparent specific mass is the average density of the specimens, which was calculated using the following equation:

$$MEA\ (\%) = \frac{Ps}{Pu - Pi} x\ 100\ (g) \qquad\qquad (11)$$

Being:

MEA = Apparent Specific Mass

Ps = dry body mass (g)

Pu = wet body mass (g)

Pi = mass of the body immersed in water (g)

3.4.8   Tension at break in bending (TRF)

Flexural stress and rupture were measured using a NANNETTI FAENZA fleximeter (Figure 10).

Figure 10 - Fleximeter (a); TRF test (b); specimen rupture (c).

The specimens were aligned on the bearings fixed to the base of the equipment, with a distance between the pieces of around 20% (twenty per cent) in relation to the length of the specimen. The device was calibrated using the number 13 programmes on the equipment.

CHAPTER 4

RESULTS AND DISCUSSION

4.1. Sample characterisation

4.1.1. Particle size analysis of clay

According to the results of the granulometric analysis (Figure 11), it can be seen that more than 80% of the clay has a granulometry of less than 0.075mm (#200 ABNT). ABNT NBR 6502/95 classifies soil particles according to the diameter of the grains, with clayey soil having a grain size of less than 0.002mm. This type of fine-grained, well-grouped soil has high plasticity characteristics, which, once wet, makes it easier to mould bricks. This statement is confirmed by Rocha (2013, p. 26), in a study carried out with soil from the same region of Iranduba/Am, A granulometric analysis of soil from the region of Pará, carried out by Jimenez (2011) observed that more than 70% of the clay had a grain size of less than 0.038mm (#400 ABNT) and these characteristics were important for incorporating sludge into ceramic mixtures.

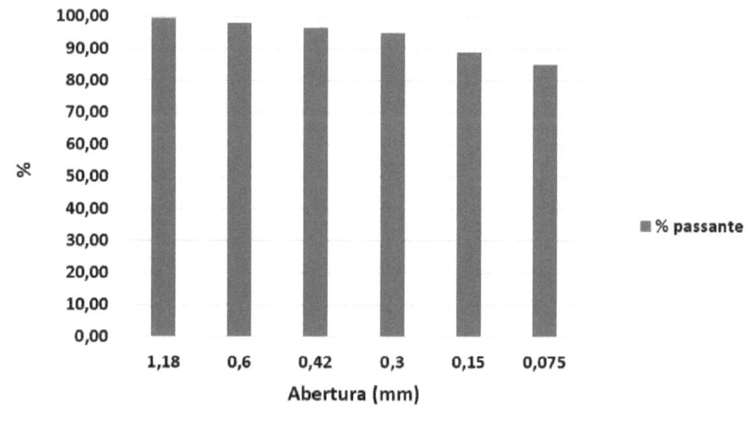

Figure 11 - Particle size distribution of clay.

4.1.2. Particle size analysis of the waste

The ABNT NBR 7181/84 standard was used to carry out the granulometric characterisation of the sludge.

It can be seen from the results of the granulometric analysis of the WTP sludge in (Figure 12), which shows the percentage of passing material according to the sieve openings, that the waste has a continuous gradation on the sieves from 0.3 to 0.075, i.e. the material analysed is well graded. Possibly without empty spaces, which prevents the material from becoming porous and increases its mechanical resistance. It can also be seen that more than 50% of the material has dimensions of less than 0.42%, which facilitates homogenisation between the grains of clay and residue.

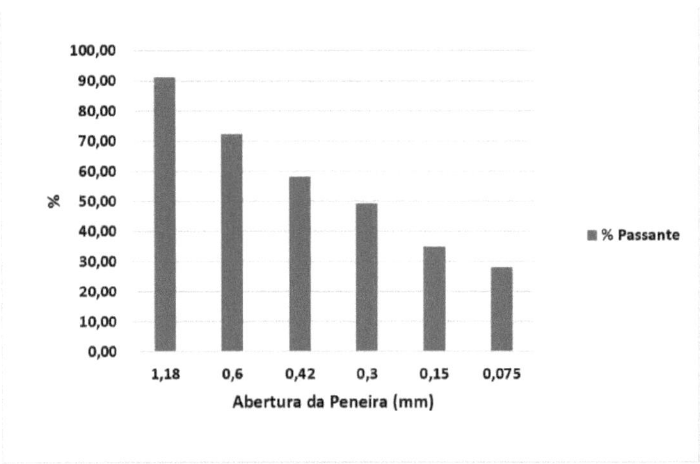

Figure 12 - Particle size distribution of the waste.

In the particle size analysis of WTP waste carried out by Silva et al (2015), it was found that the particle size distribution was in the range of 0.001 to 1.2mm, these results showed a wide range of particle size distribution, as they chose to use the waste that came out of the grinder without prior classification of the particle size. Silva el al (2015) also found that the sludge particles had an irregular profile due to the fact that they had not obtained a particle size classification.

standard granulometry in the sludge particle preparation process. In this study, the well-graded and continuous format of the WTP sludge residue is considered a positive result, because with the regular decreasing graduation of grain sizes, it facilitates the homogenisation and aggregation of the soil particles during cooking.

4.1. 3 Actual grain   density (DR)

The results of the real density of the clay and silt grains (Table 4) reflect the presence of minerals and organic matter in the soil. The real density limits vary between 2.3 and 2.9 g/cm^3 , it can be seen in Table 4 that both the clay and the residue have a low RD, between 1.717 and 1.718 respectively, which characterises a soil with high organic matter, positively influencing porosity, as few porous cavities do not weaken the ceramic material.

Table 4 - Actual grain density of clay and silt samples.

Standard: DNER-ME 093/94	Clay	Sludge	Standard: DNER-ME 093/94	Clay	Sludge
Pycnometer	1	2	Weight of pycnometer + sample + water (g)	159,55	160,01
Temperature °C	25	26	Weight of pycnometer + water (g)	151,19	152,20
Pycnometer weight (g)	47.46	47,46	Weight of dry material (g)	20,0	20,0
Weight of pycnometer + dry sample (g)	67, 47	66,15	Real grain density (g/cm)^3	1,717	1,718

4.1.4. X-Ray Fluorescence

The X-ray fluorescence results for both the clay and the residue (sludge) showed the highest concentrations of aluminium oxide (A12O$_3$) at 51.33% and silicon oxide (SiO$_2$) at 55.41 (Table 5). Silicon oxide is largely present at the base of the structure of clay minerals, so the presence of this compound in the materials analysed was expected. As for A12O3, it is mainly due to the type of coagulant (aluminium sulphate) used during water purification at the PROAMA Treatment Plant. This fact was found by Ramos (2015), in analyses of the elements in sludge and clay, where he found a higher concentration of Aluminium Oxide (A1203) and Silicon Oxide (SiO$_2$), due to the type of coagulant used in decantation and the mineralogical sedimentation characteristics of the Amazon region. Barbosa (2000) found in his studies that the predominant elements in the sludge were aluminium, iron and silica; the highest content was aluminium, due to the use of aluminium sulphate as a coagulant. This chemical similarity between the materials corroborates the addition of sludge to the clay mix for making ceramic bricks, as it avoids unwanted reactions during the firing process of the specimens. In tests, Ramos (2015) proved that incorporating 10% sludge into the ceramic mass was physically and mechanically efficient for making bricks, due to the similarity between the materials.

Table 5 - Chemical composition in percentage of clay, residue and ceramic mass with 20% residue.

| | X-Ray Fluorescence | | |
| | Concentration (%) | | |
Compound	Clay	Waste	Ceramic clay with 20% residue
Al_2O_3	51, 33	55,41	51,44
CaO	0,01	3,95	0,17
Cl	0,006	0,04	0,006
Cr_2O_3	0,008	-	-
Fe_2O_3	2,71	6,61	3,34
K_2O	0,37	1,25	0,45
MgO	0,18	0,48	0,21
MnO	0,005	0,03	-
$On O_2$	0,14	0,31	0,22
Nb_2O_5	0,003	0,02	0,004
$P2O_2$	0,06	0,31	0,05
$Rb O_2$	0,002	-	-
SiO_2	45, 74	36,38	42,21
SO_3	0,05	1,44	0,09

4.1.5 X-ray diffraction

The results of the microstructural characterisation of crystalline materials carried out using X-ray diffractometry on the clay are shown in Figure

13. It is possible to observe that the most accentuated phases are: quartz (SiO_2) with the majority peak, followed by kaolinite ($Al_2 Si_2O_5 (OH)_4$). This is due to the fact that most of the clay minerals in the state of Amazonas come from residual deposits formed from alterations to rocks of the Alter do Chão Formation composed, among others, of quartz and kaolinite. According to Campos (2008), most of the region around the Manacapuru-Iranduba ceramics centre is made up of these residual deposits. According to Jimenez (2011), in

analyses of clay extracted from the banks of the Guamá River in Pará, quartz (the majority), followed by kaolinite, was also found in that region. The quartz mineral found in this study confirms the chemical composition found in the X-ray fluorescence analysis, since the chemical composition of quartz is silica dioxide. Quartz influences thermal expansion during cooking

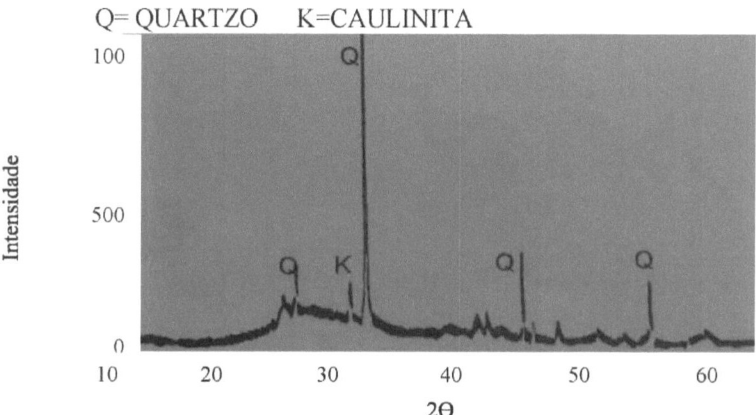

Figure 13 - X-ray diffractogram of the clay. Source: author, 2015.

It can be seen that the most intense mineral found in the residue, as in the clay analysis, is quartz, followed by kaolinite (Figure 14).

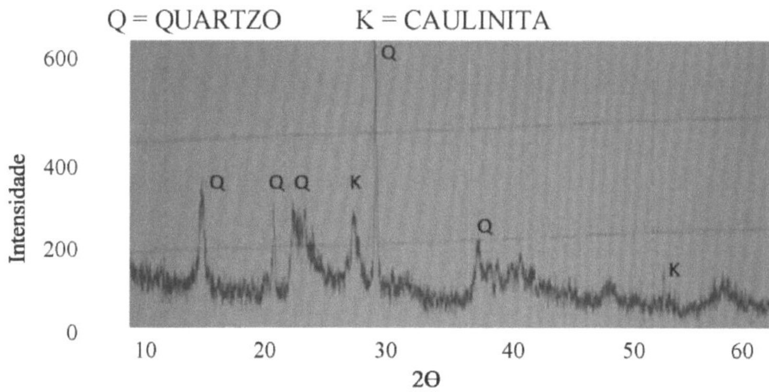

Figure 14 - X-ray diffractogram of WTP residue. Source: author, 2015.

Silva (2008) states that this similarity of minerals between clay and sludge is extremely important due to the physical and mechanical benefits of incorporating the waste into the clay

31

mass.

Figure 15 shows the results of the X-ray diffractograms of the specimens with a proportion of 20% waste, after firing at 900°C. It can be seen that there was a favourable increase in quartz peaks and a reduction in kaolinite intensity. Ramos (2015) argues that the reduction in kaolinite was due to the sintering of the specimens, with the linear organisation of the minerals present, improving the quality of the final product.

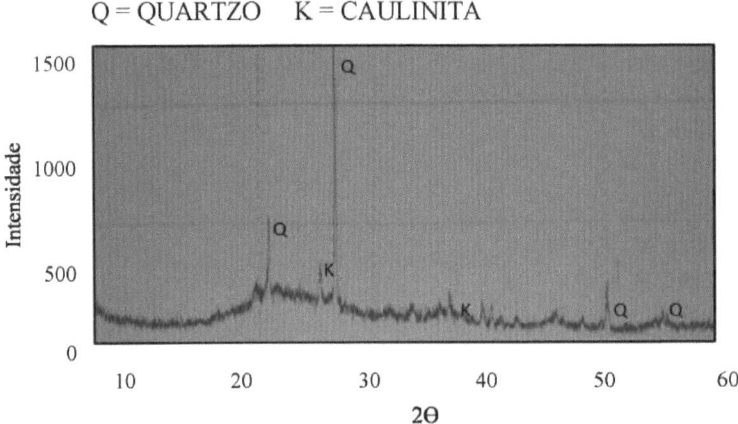

Figure 15 - X-ray diffractogram of the CP after firing, with 20% residue. Source: author, 2015.

Jimenez (2011) reinforces that this behaviour occurs due to the addition of residue which favours the appearance of hematite and the more residue added, the iron in the sludge tends to accelerate the solid state reactions, giving rise to a new crystalline phase, cristobolite. Tartari el al (2011), in an analysis of clays exploited in the western region of Paraná, observed that the clays contained three main minerals: kaolinite, a hydrated aluminium silicate mineral, quartz, a silicon oxide mineral, and hematite, an iron oxide mineral.

4.1.6   Results of the physical and mechanical analyses

4.1.6.1   Drying of the specimens

It can be seen from the results of the average dimensions and weight of the specimens (Figure 6) that, after drying, there was a contraction in all the dimensions of the samples, showing that there was an elimination of moisture which favours better mechanical performance of the product. Despite the shrinkage in the volume of the specimens, no defects were observed in the samples. Studies carried out by Ramos (2015) showed that after drying the specimens at 110°C, there was a volume contraction in all dimensions.

Table 6 - Test specimen measurements after drying at 110°C.

Specimens	Ceramic putty	Dimensions after drying (mm)			Weight (g)
		Length	Width	Thickness	
1	0%	72,11	31,49	7,70	26,87
2	12%	72,06	31,49	7,82	26,98
3	15%	72,09	31,52	7,89	26,88
4	20%	72,08	31,49	7,92	26,93
Averages		**72,08**	**31,49**	**7,83**	**26,91**

4.1.6.2   Firing curve

The muffle furnace was programmed to raise the final temperature to 900 °C in 2h:30min of firing, as well as remaining on the first two levels for 5 minutes and on the last for 0.30 minutes. The red line represents the ideal temperature programmed into the muffle furnace and the green line, the actual firing temperature, where a more marked variation of around 120°C can be seen (Figure 16). The results show that there was little change in temperature during firing, resulting in the pieces being baked in a controlled manner with good results, as the specimens did not deform or crack. During the firing of the specimens, it is important to control the temperature to avoid sudden water loss which will cause excessive shrinkage and consequently cracking (HENTGES, 2012, p. 15).

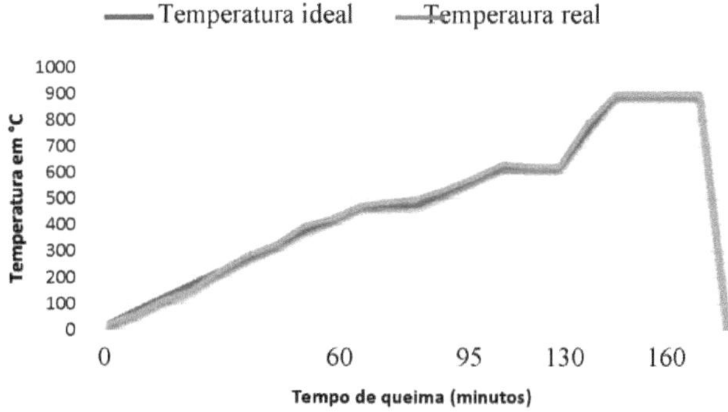

Figure 16 - Firing curve of the test specimens. Source: author, 2015.

4.1.6.3   Physical and mechanical   analyses of the specimens after firing at 900°C.

Table 7 shows the average physical and mechanical results of the linear shrinkage (RL), water absorption (AA), apparent porosity (PA), apparent specific mass (MEA), tensile strength (TR) and flexural tensile strength (TRF) tests carried out on the specimens after firing at 900°C, with 4 concentrations of residue: 0%, 12%, 15% and 20%. 5 specimens were produced for each formulation. Based on these results, Figures 17, 18, 19, 20 and 21 were drawn up to analyse each result.

Table 7 - Results of the physical and mechanical tests.

Formulation			Firing temperature at 900°C				
Waste	Clay	Water	RL (%)	AA (%)	AP (%)	MEA (%)	TRF (Kgf/mm)^2
0%	90	10%	3,30	16,21	21.00	1,77	2,41
12%	78	10%	3,38	11,62	19,94	1,68	2,37
15%	75	10%	3,39	11,20	13,76	1,69	2,20
20%	70	10%	3,60	8,73	15,32	1,73	1,80

4.1.4.1.   Linear Shrinkage (LR)

It can be seen from the linear shrinkage results after firing at a temperature of around 900°C, depending on the percentages of 0%, 12%, 15% and 20% of waste added to the clay mix (Figure 17) that linear shrinkage increased with the increase in the sludge content of the clay mix in all percentages, which is due to the sintering process of the specimens heated to high temperatures. The grouping of molecules is an important and necessary factor in the firing of ceramic materials, as it causes hardening to occur and consequently gives the final product strength. Callister (2012) reinforces that during firing the pieces contract due to the coalescence of the powder particles by the sintering process. Askeland (2011, p. 455) states that the driving force behind sintering is the reduction in the surface area of the powder. Jimenez (2001, p. 74) points out that this behaviour indicates an improvement in the sintering process of the specimens. This sintering process, which takes place during firing, creates a change in the microscopic structure due to minerals such as quartz, making the piece solid, which results in the ceramic product having good resistance.

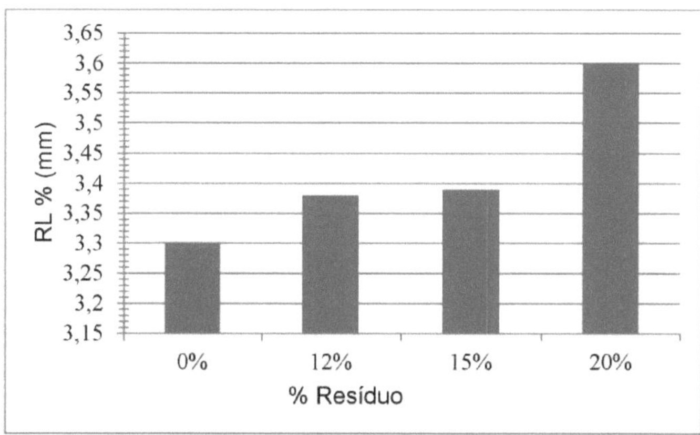

Figure 17 - Linear shrinkage in the test specimens.

Ramos (2015) considers that the incorporation of WTP sludge into the ceramic brick manufacturing process, in the proportions of 0%, 3%, 5% and 10% of waste, resulted in an increase in linear shrinkage at a sintering temperature of 900°C, possibly due to the presence of organic matter that volatises during firing. Jimenez (2001) stresses that organic matter volatises when ceramic material is fired at high temperatures.

According to ABNT (NBR 15270-3, 2005), all the shrinkage values comply with the limits set by the standard, which is below 8%.

4.1.6.5.   Water absorption

The results of the water absorption test of the specimens according to the percentages of waste added to the clay - 0%, 12%, 15% and 20% - can be seen in Figure 18. It can be seen that the greater the amount of waste incorporated into the clay mass, the lower the occurrence of water absorption. According to Callister (2012, p. 441), each of the raw materials influences the changes that occur during the firing process and the characteristics of the finished piece. In this case, water absorption decreases as the silt content in the clay mixture increases in all proportions, confirming that the more sintered the material is depending on the firing temperature, the greater the linear shrinkage and the lower the water absorption due to the grouping of the molecules. Ramos (2015, p. 18) noted that with an increase in residue there was a decrease in water absorption. Jimenez (2011, p. 75) states that the decrease in AA is expected due to the large presence of clay and low presence of sand. Almeida (2015) emphasises that organic matter and humidity are important parameters that can influence the final quality of ceramic blocks.

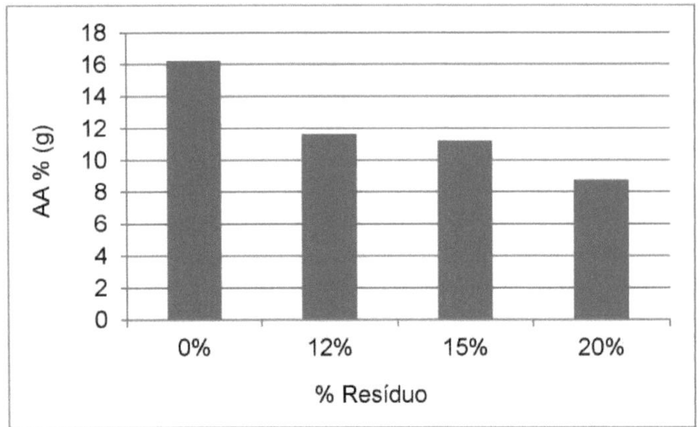

Figure 18 - Water absorption of the specimens.

Tsutiya (2001, p. 05) states that humidity is important in determining the handling of sludge, as a high humidity content can jeopardise the routing of manufacturing components, obstructing passages or adhering to parts of the system. According to ABNT (NBR 15270-3, 2005), the maximum and minimum water limits are 8% to 22%, respectively. It should be noted that the moisture content results are within the limits allowed by the standard, implying a good result for the material studied.

4.1.6.6 Apparent porosity (AP)

Callister (2012, p. 417) argues that after compacting the powder particles into the desired shape, there will be pores or empty spaces between the particles and that after heat treatment most of this porosity will be eliminated, but some residual porosity will remain. Figure 19 shows the results of the average apparent porosity after firing at 900°C, depending on the formulations with 0%, 12%, 15% and 20% waste added to the ceramic mass. It can be seen that at a temperature of 900°C, there is a contraction in the specimens causing a reduction in porosity. Callister (20012) states that these microstructural changes of shrinkage and pore reduction occur during firing.

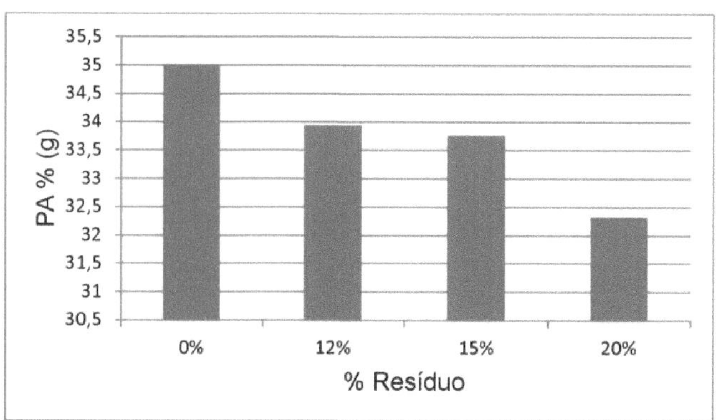

Figure 19 - Apparent porosity of the specimens.

Along the same lines, Callister (2012, p. 446) states that during baking, the moulded part contracts and shows a reduction in its porosity, along with an improvement in its mechanical integrity. Jimenez (2011, p. 76) confirms that apparent porosity decreases as the temperature rises. Another aspect raised by Jimenez (2011, p. 76) is that the reduction in water absorption is similar to the reduction in porosity. The same behaviour was observed by Ramos (2015, p. 19), whose tests showed a reduction in absorption and porosity, noting that there was no discrepancy in relation to the proportions studied of 0%, 5%, 7% and 10%.

For this study, the percentage of porosity in the pieces is very important, because the greater the porosity and water absorption, the lower the strength and quality of the ceramic brick, in addition to other advantages. Ramos (2015) states that low porosity means greater thermal comfort and less possibility of infiltration for use in the construction and structuring process. However, Askeland (2008, p. 460) argues that total porosity negatively affects the mechanical properties of ceramics; on the other hand, he points out that porosity can be useful for increasing resistance to thermal shock. Along the same lines, Callister (2012, p. 418) shows that porosity has a negative effect on flexural strength due to the reduction in porosity. It is possible to assume that, due to these characteristics of low porosity, the proportions of 0%, 12%, 15% and 20% of waste incorporated into the ceramic mass will not compromise flexural strength, which is why the aim of this study is to successfully incorporate sludge into clay for brick manufacture, since what is expected is a material that is not resistant to shock, but rather a product that supports and resists the tension applied to it, in accordance with standard 15270-1,2,3 (2005).

4.1.6.7.   Apparent specific mass (ASM)

The results of the apparent specific mass of the ceramic products were analysed as a
function of firing temperature for the formulations with 0%, 12%, 15% and 20% waste
incorporated into the clay mass (Figure 20). It can be seen that at 900°C there was an increase
in MEA in all the formulations. This was also shown by Souza (2010), whose MEA increased
under the influence of high temperatures. It can be seen that the MEA of the proportions was
between 1.85% and 2%. Ramos (2015) states that this small difference between the values
demonstrates the good performance of the residue added to the clay mix. Ramos also explains
that the increase in mass is due to the density of the incorporated sludge, as well as the organic
matter that dissipates in the process of firing the specimens.

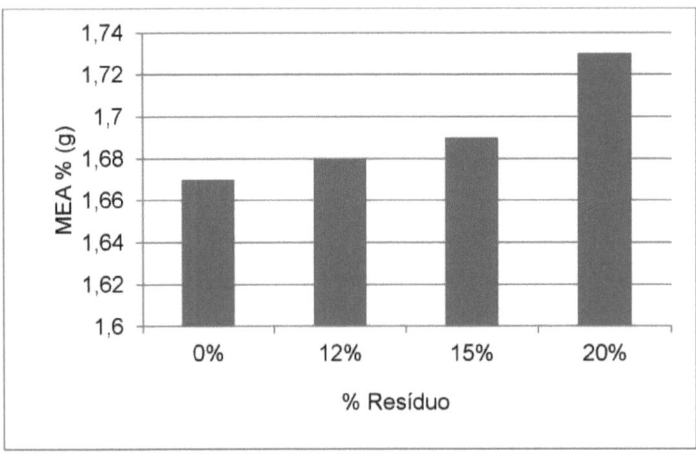

Figure 20 - Apparent specific mass of the specimens.

Aquino (2015, p. 32) found a decrease in the specific mass as the waste content increased.
This was due to the addition of up to 25g of textile industry sludge waste in the production of
ceramic sealing blocks at a high temperature of 1150°C. In this study, WTP sludge was mixed
into the ceramic mass at a maximum temperature of 900°C, resulting in an increase in mass
and a decrease in water absorption and apparent porosity. These characteristics are to be
expected, as they positively favour the strength of the final product. The increase in the mass

and volume of the specimens was also observed by Jimenez (2011) when adding up to 10% of residue to the clay mass, which is due to the unconventional behaviour of kaolinite raw materials used in red ceramics. Teixeira (2006) states that apparent porosity and apparent specific mass are associated with water absorption.

4.1.6.8.   Flexural tensile strength (FRT)

It can be seen that the averages of the flexural tensile strength (FRS) results as a function of the 900°C firing temperature of the specimens in the 0%, 12%, 15% and 20% formulations (Figure 21) showed a decrease in FRS with the incorporation of the residue, The linear shrinkage, water absorption and apparent specific mass tests confirmed this with the tensile strength test, as the values found are within the range recommended by standard 15270-3 (2005). This result was also found by Ramos (2015), in studies on the addition of up to 10% sludge, and he states that this decrease in TRF was expected, due to the addition of residue to the ceramic mass, however it does not negatively influence the specimens. Teixeira et al (2006) in tests showed that the TRF decreased with the concentration of sludge added to the ceramic mass, confirming a good result for the material.

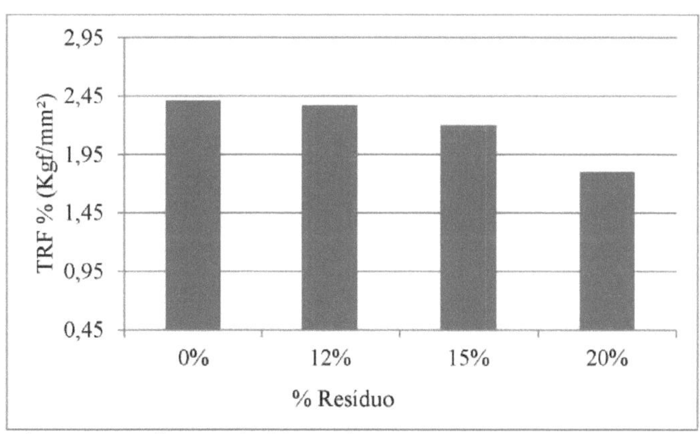

Figure 21 - Tensile strength of the specimens.

CHAPTER 5

CONCLUSION

The granulometric characterisation of the soil samples showed that the clay has more than 80% granulometry of less than 0.075mm (#200 ABNT) and is classified as clay soil.

As for the sludge residue, according to the granulometry data, it is well graded and continuous, in accordance with ABNT NBR 7181/82.

In the chemical analysis by X-ray fluorescence, both in the clay and in the WTP sludge residue, the elements found in highest concentration were Aluminium Oxide (Al_2O_3) and Silicon Oxide (SiO_2), due to the type of coagulant used in decantation and the mineralogical sedimentation characteristics of the Amazon region.

The physical and mechanical analyses, such as water absorption, apparent porosity, apparent specific mass and flexural tensile strength, show that WTP sludge in proportions of no more than 20% can be incorporated into clay for brick manufacture with satisfactory results, as they comply with the parameters established by ABNT (NBR 15270-1,2 and 3 of 2005).

Contributing to a reduction in the mineral extraction of clay, as well as a reduction in the dumping of sludge in rubbish dumps or bodies of water, will bring benefits not only to the environment by reducing the impacts generated by potteries and water treatment plants, but also to society as a whole.

REFERENCES

Regulatory Agency for Concessioned Services in the State of Amazonas. Activities Report. 2010.

ALMEIDA. P.H.S. et al. **Influence of the type of clay in the textile sludge solidification/stabilisation process.** Ceramics 61, 137-144, 2015.

ALVES, Diego Henrique Correa, et al. **Solid Waste, more than an Environmental Issue, a Social Issue.** Engineering and Science, 2015.

ANDRADE, P. S. **Environmental impact assessment of the use of water treatment plant**

waste in the red ceramics industry: a case study. SP, 2005.

AQUINO, R.C, et al. **Addition of textile industry sludge waste in the production of ceramic sealing blocks.** Vol 10. N. 1. Electronic Journal of Materials and Processes, p. 29-35. 2015.

ASKELAND, D.R. Book: Materials Science. Trilha. São Paulo. 2014.

BRAZILIAN ASSOCIATION OF TECHNICAL STANDARDS. **NBR 15.270-1**: Ceramic components - Part 1: Ceramic blocks for structural masonry - Terminology and requirements. Rio de Janeiro, 2005.

BRAZILIAN ASSOCIATION OF TECHNICAL STANDARDS. **NBR 15.270-2**: Ceramic components - Part 2: Ceramic blocks for sealing masonry - Terminology and requirements. Rio de Janeiro, 2005.

BRAZILIAN ASSOCIATION OF TECHNICAL STANDARDS. **NBR 15.270-3**: Ceramic blocks for structural and sealing masonry - Test methods. Rio de Janeiro, 2005.

BRAZILIAN ASSOCIATION OF TECHNICAL STANDARDS. **NBR 6457:** Soil sample - Preparation for compaction tests and characterisation tests. Rio de Janeiro, 1986.

BRAZILIAN ASSOCIATION OF TECHNICAL STANDARDS. **NBR 6502:** Rocks and soil. Rio de Janeiro, 1995.

BRAZILIAN ASSOCIATION OF TECHNICAL STANDARDS. **NBR 7181:** Soil - Particle size analysis - Test method. Rio de Janeiro, 1984.

BRAZILIAN ASSOCIATION OF TECHNICAL STANDARDS. **NBR ISO 10004:** Solid waste: Classification. Rio de Janeiro 2004.

BRAZILIAN ASSOCIATION OF TECHNICAL STANDARDS. **NBR NM-ISSO 2395**: Test sieves and sieving tests-Vocabulary. 1997.

BARBOSA, R.M.; et al. **Toxicity of water treatment plant sludge to daphniasimilis (ceadocera, crustacea).** International Congress of Sanitary Engineering, XXVII; Porto Alegre Aidis, p. 10. 2000.

BARROSO, M. M.; CORDEIRO, J. S. Metals and Solids: **Legal Aspects of Water Treatment Plant Waste**. In: Brazilian Congress of Sanitary and Environmental Engineering. Proceedings... João Pessoa: ABES, 2001 (Published on 1 CD-ROM).

BILDHAUER, D.C. et al. Solid bricks with refractory characteristics from the incorporation of marble and granite waste. Porto Alegre. 2015.

BOTERO, W. G. **Characterisation of sludge generated in water treatment plants: prospects for agricultural application.** São Paulo. Quim. *Nova*, Vol. 32, No. 8, 2018-2022, 2009.

BRAZIL. MINISTRY OF THE ENVIRONMENT. **Environmental Crimes.** Law No. 9.433 of 8 January 1997. Brasília. http: //www.mma.gov. br/ Accessed on 10 September 2015 at 11:30.

BRAZIL. MINISTRY OF THE ENVIRONMENT. **National Water Resources Policy.** Law No. 9.605 of 12 February 1998. Brasília. http://www.mma.gov.br/ Accessed on 10 September 2015 at 11:00.

BRAZIL. MINISTRY OF THE ENVIRONMENT. **National Solid Waste Policy.** Law 12.305 of 3 August 2010. Brasília. http://www.mma.gov.br/ Accessed on 10 September 2015 at 10:00.

CALLISTER, W. D. and RETHWISH, D.G. **Materials science and engineering: an introduction**. Rio de Janeiro: LTC, 2012.

CAMPELO, N.S. et al. **Study of the use of burnt ceramic residue ("chamote") from the pottery cluster in the municipalities of Iranduba and Manacapuru, Amazonas, as an additive in the manufacture of roof tiles.** Cerâmica Industrial, p. 11. Amazonas. 2006

CAMPOS, A.M.L.S. **The effect of firing temperature on calcined clay synthetic aggregate applied to asphalt concrete**. Dissertation presented to the Postgraduate Programme in Civil Engineering at the Federal University of Amazonas. Manaus, 2008.

CASTRO, T. M. et al. **Characterisation of acoustic ceramic blocks produced with the incorporation of textile laundry sludge**. Eng Sanit Ambient | v.20 n.1 | jan/mar 2015.

COELHO, A.C.V. and SANTOS, P.S. **Special clays: what they are, characterisation and properties.** São Paulo-SP. Química Nova. P. 146-152, 2007.

PARANÁ SANITATION COMPANY - SANEPAR. **Diagnosis of sludge from water treatment plants in the state of Paraná**. Sanepar, Curitiba, 2013.

GIL, A. C. Como elaborar projetos de pesquisa. 4.ed. São Paulo: Atlas, 2002.

HOPPEN, C. et al. **Co-disposition of centrifuged Water Treatment Plant (WTP) sludge in concrete matrix: alternative method for environmental preservation**. Cerâmica 51. p. 85-95. 2005.

JIMENEZ, I,T. **Utilisation of Plastic Injection Industry Effluent Treatment Plant Sludge as a Raw Material for the Ceramic Industry**. Dissertation (Master's Degree in Electrical Engineering) - Postgraduate Programme in Electrical Engineering. Federal University of Pará, ITEC Technology Centre. 2011.

KATAYAMA, V. T. **Quantification of sludge production in full cycle water treatment plants: a critical analysis**. São Paulo, p. 139. 2012.

KOPEZINSKI, Isaac. **Mining x the environment: legal considerations, main environmental impacts and their modifying processes**. Porto Alegre: Ed. Universidade/UFRGS, 2000.

PINHEIRO, B. C. A.; ESTEVÃO, G. M.; SOUZA, D. P. **Sludge from the water treatment plant in the municipality of Leopoldina, MG, for use in the red ceramics industry Part I: characterisation of the sludge**. p.204-211, 2014.
PINTO, N. M. M.; PINHEIRO, H. A. **Os impasses da Legislação Ambiental para atividade oleira em Iranduba (AM): Entre a Lei e os danos.** Proceedings of the Brazilian Congress on Environmental Management and Sustainability. Vol 1, Congestas p. 434, 2013.

AMAZONAS MINERAL CLUSTER. **The Mineral Production Centre**. Giz *Deutsche Gesellschaft Fur Internationale Zusammenarbeit (GIZ) GmbH*. 2014.

PORTELA, M.O.B and GOMES, J.M.A. **The environmental damage resulting from clay extraction in the potteries neighbourhood of Teresina, PI**. São Luís- MA, August 2005.

RAMOS. S.B.J. **Incorporation of Water Treatment Plant Sludge into the Ceramic Brick Manufacturing Process**. Final Coursework for the Environmental Engineering Degree at the University Centre of the North - Uninorte/Laurreate. 2015.

REIS, E.L.T. et al. **Evaluation of the environmental impact of water treatment plants on watercourses.** 2006.

ROCHA, N. F. **The chemistry of clays and ceramics - an approach for secondary education. Course Conclusion Programme**. University of Brasília Institute of Chemistry. Brasília, 2013.

SILVA, C.A. et al. **Incorporation of water treatment sludge in the manufacture of chipboard panels**. Engevista, V. 17, n3, p. 398-406, 2015.

SILVA, J. F.A. **Behaviour of asphalt concrete with sludge from the Manaus Water Treatment Plant as filler. manaus** - Dissertation (Master's Degree in Civil Engineering) Universidade Federal do Amazonas, 2008. Manaus: UFAM, 2008.

SILVA, N.A. et al. **Evaluation of environmental performance in a ceramic industry company in Tocantins**. Revista Eletrónica em Gestão, Educação e Tecnologia Ambiental Santa Maria, v. 19, p. 848-861. 2015.

SILVA. R. L. et al. **Feasibility of incorporating sewage treatment plant (STP) sludge into ceramic mass for block production**. Cerâmica 61. p. 3140. 2013.

SOARES, L. A. et al. **Characterisation of WTP waste for disposal at the outlet of Stabilisation Ponds.** 2014.
SOUZA, F.R. Compost of water treatment plant sludge and wood sawdust for use as coarse aggregate in concrete. Thesis (Doctorate in Materials Science and Engineering). University of São Paulo, São Carlos. 2010.

TAKADA, C.R.S. et *al.* **Utilisation and final disposal of sludge from water treatment plants in the municipality of palmas - to.** p. 157-165. 2013.

TARTARI, R. et al. **Sludge generated at the Tamanduá water treatment plant, Foz do Iguaçu, PR, as an additive in clay for red ceramics. Part I: Characterisation of sludge and clay from the third plateau of Pará**. Cerâmica 57, p 288-293, 2011.

TEIXEIRA, S. R. et al. **Characterisation of water treatment plant (WTP) and sewage waste and the feasibility of its use by the ceramics industry.** XXVIII International Congress of Sanitary and Environmental Engineering. Mexico 2002.

TSUTIYA, M. T.; HIRATA, A. Y. **Utilisation and final disposal of sludge from water treatment plants in the State of São Paulo**. In: 21st Brazilian Congress of Sanitary and Environmental Engineering, ABES, João Pessoa, 2001.

Printed by Books on Demand GmbH, Norderstedt / Germany